Benjamin Brodie

Ideal Chemistry

A Lecture

Benjamin Brodie

Ideal Chemistry
A Lecture

ISBN/EAN: 9783337026707

Printed in Europe, USA, Canada, Australia, Japan

Cover: Foto ©berggeist007 / pixelio.de

More available books at **www.hansebooks.com**

IDEAL CHEMISTRY.

A Lecture.

BY

Sir B. C. BRODIE, Bart., D.C.L., F.R.S.,

Professor of Chemistry in the University of Oxford.

$$2a\chi = a + a\chi^2$$

London:

MACMILLAN AND CO.

1880.

PREFACE.

THE following Lecture was delivered before the Chemical Society on June 6, 1867, after the presentation to the Royal Society of my first Memoir on the Cálculus of Chemical Operations. The Lecture, however, has not been published except in a report which appeared in the *Chemical News* of June 14, 1867. This report I have in the main followed. It is, however, far from presenting a satisfactory account of the Lecture; and indeed, in several important passages entirely fails to represent my meaning. I publish ·this Lecture now, partly that the views given. in it may be correctly apprehended, and also that I think it will have a wider interest, and be more generally appreciated by those who are curious in these

questions than at the time it was delivered,
when the whole subject was new, and imper-
fectly understood.   Also, although the Lecture
is short, it touches upon two or three topics
of fundamental importance, which I have not
elsewhere discussed in the same way.   Of these,
there are three which I may especially indicate.
Firstly, the application which I have made of
the symbol xy regarded as a chemical symbol;
secondly, the meaning to be assigned to the
term "ideal element," and lastly, the suggestion
which is here made, I believe for the first
time (excepting in the few words at the conclusion of Part I. of the Memoir referred to
above), of the possible decomposition, at the
elevated temperature of the sun, of certain
chemical elements, and of the existence in
that luminary of their constituents in independent forms.

February 26, 1880.

IDEAL CHEMISTRY.

A Lecture delivered before the Chemical Society, on Thursday, June 6th, 1867.

MR. PRESIDENT,—I feel that I have undertaken this evening a truly difficult task, to give to the Chemical Society, in the brief space of one hour, an account of an abstruse and difficult subject, the exact comprehension of which requires that it should be minutely considered in all its details. I should not, however, shrink from this, if I did not feel that the subject is really before those even who are competent to judge of it, in a somewhat imperfect form ; that I have as yet offered to the chemical world the first part only of the method of which I am about to speak ; and that this method will be much better comprehended, both from a mathematical and chemical point of view, when you have before you the subsequent parts which I hope to present hereafter.

I am to speak of a method of representing the facts of chemistry, which is fundamentally different from the method at present in use. Let me say a few words upon the past history of chemical theories.

I believe theory to be essential to the existence of chemistry. The birth of the science was inaugurated by the construction of a definite theory of chemistry—the first theory which had ever been proposed, and which sought to give a definite and rational account of the facts of the science. Th;s theory was the once world-famous doctrine of Phlogiston. In this theory the facts of chemistry were explained by the agency of a subtle, all-pervading, hypothetical principle, by the transference of which, from one chemical substance to another, it was assumed that the facts of chemistry were correctly accounted for. It is easy, from our present point of view, to pass critical remarks upon the doctrine of Phlogiston, but it is not quite so easy really to comprehend that doctrine and to put ourselves in the position of those great chemists who worked and who studied through its agency. If ever any one be tempted to speak slightingly of the doctrine of Phlogiston, let him remember that through the instrumentality of this doctrine the great discoverer of chlorine, the chemist Scheele, worked. Let him remember that the exact mind of Cavendish was contented with this doctrine. Let him remember again that the illustrious Priestley, that transcendentally inventive genius, in possession of this doctrine, made the great discovery of oxygen: and that not only was he then contented with this theory but that he died a firm believer in and adherent to it. However, the doctrine of Phlogiston, like many human surmises, was destined to pass away—Lavoisier shattered Phlogiston. For no inconsiderable period after this chemists appear to have worked, if I may so say, without a theory; that is to say, that, as during the long alchemical ages chemists were occupied in collecting together those facts which

were afterwards to be embodied in the theory of
Phlogiston ; so for a period of above thirty or
forty years—that is to say, from the time of
Lavoisier to the time of Dalton—chemists were
employed in collecting together that exacter
system of facts which was to form the basis of
a far wider, and far more comprehensive theory,
namely, the great atomic doctrine.   However,
Davy appears to have worked and to have made
his great discoveries without a theory.   Davy
never admitted the atomic theory, but rested
content simply with the facts of numerical
analysis and the laws of combination deduced
from them.

In the year 1808 appeared that famous book,
A New System of Chemical Philosophy, which
contained the germs—indeed, I may say, almost
the full development—of the atomic theory itself.
In this atomic doctrine Dalton took up the
conception of combination, which was introduced
into the science by means of the theory of
Phlogiston.   He took up that doctrine of combination, and moulded it into a new and more

definite form. It would be useless for me, before the Chemical Society, to dwell upon the atomic theory. It is a theory with which every one is familiar, for every chemist of this day has worked with that theory, has conceived his science from the points of view of that theory; and, indeed, I believe, in the opinion of many, it is almost impossible that that doctrine should ever fall to the ground. This doctrine of Dalton, however, was a doctrine far more audacious than that of Stahl. In the theory of Phlogiston, Stahl at least considered that he had visible and palpable evidence of the transference of his Phlogiston from chemical system to chemical system; but Dalton told us that this notion of the continuity of matter—that obvious fact which our senses teach us—was simply an illusion, and that, if only we could see things aright, we should see that this world, which appears to us so connected and so continuous, was really made up of disjointed fragments.

From the point of view of the atomic theory, chemists have worked for a period now of about

sixty years, and the progress of chemical theory
has consisted in the almost constant and unremitting development of this doctrine. I cannot
say, however, that this has been an unremitting
progress. It has rather been a succession of
changes. System has followed system, doctrine
has followed doctrine ; but these doctrines have,
one after another, fallen to the ground. We
have had but little that is permanent, and at
the present moment the theory of chemistry is
built upon the ruin of other theories. Now no
one can have more respect for these great ideas
which were thus ushered into the science by
Dalton, than I myself have. It cannot be necessary for me to express to this Society of
Chemists the admiration which I as a chemist
feel for that theory ; but, nevertheless, it is no
disparagement to say that I think the atomic
doctrine has proved itself unable to deal with
the complicated system of chemical facts, which
has been brought to light by the efforts of
modern chemists, and has not succeeded in constructing an adequate, a worthy, or even a

thoroughly useful representation of those facts, although for sixty years the united efforts of chemists, including many of the most able men in science, have been devoted to the development of this doctrine, and have founded their representations upon it. Now, let me read to you an account of the last modern representation of the atomic doctrine, and the chemical symbols in which the atomic doctrine has resulted. I will read to you a paragraph headed "Glyptic Formulæ;" it is given in a scientific journal, *The Laboratory.* Here is the paragraph :—

"Those teachers who think, with Dr. Frankland and Dr. Crum Brown, that the fundamental facts of chemical combination may be advantageously symbolized by balls and wires, and those practical students who require tangible demonstration of such facts, will learn with pleasure that a set of models for the construction of glyptic formulæ may now be obtained for a comparatively small sum. At first sight the collection of bright-coloured and silvered balls suggests anything but abstract chemical truth." . . . .

And so on. The writer proceeds to inform us what we may procure for our money:—

"There are seventy balls in all for the representation of atoms—monads, dyads, triads, tetrads, pentads, and hexads, being distinguished by the number of holes pierced in the balls. To connect these into rational formulæ"—[which I confess I should imagine to be a truly difficult problem]—"brass rods, straight or bent, and occasionally flexible bands, are employed."

However, the editor seems to have had some misgivings, for he proceeds to say:—

"Whether they are calculated to induce erroneous conceptions is a question about which much might be said." Now, however much might be said upon this subject, I certainly am not going to say a great deal to the Society about it; but it is truly a remarkable fact, that the atomic theory, after so many efforts at completion should have resulted in such a thoroughly materialistic bit of joiner's work as this. Indeed I cannot but say that the promulgation of such ideas—even the partial reception of such views—indicates that the science must have got, somehow or another, upon a wrong track ; that the science of chemistry I say must have got, in its modes of representation, off the rails of philosophy, for it really could only be a long series of errors and of misconceptions which could have landed us in such a bathos as this.

You may, however, ask me, and with reason, " In what way, then, are we to represent the facts of chemistry, if we are not to represent them in this way ? Do you mean to deal with this complicated system of facts, and to offer us no mode of representing these facts, and no mode of conceiving these facts ?" Now, I quite admit that any person who seriously attacks these ideas, is bound to show some other, and, even some better way of representing the facts. He is bound to do this, or to refrain from his attacks. You ask me, how are we to represent the facts of the science ? It is to that question that I wish to offer an answer to-night. I say that we are to express the numerical facts

of the science by means of symbols; but I
attach to the term "symbol" a very special
signification. We have plenty of what are called
"chemical symbols" already; but these chemical
symbols are not, from my point of view, symbols
at all, and you will presently see why. Now a
symbol may be regarded as a mark by which
we express the objects of our thoughts for the
purpose of reasoning about those objects; and
one which is capable of being combined with
other similar marks according to certain definite
laws of combination; which laws of combination
are to be possible, through the interpretation of
the symbol, in the subject matter which is
symbolized. That is what I mean by a symbol.

You will readily see that our present notation
really can hardly be called, even in courtesy, a
symbolic representation. The reason is, in the
first place, that these letters H, O, &c. are not
capable of being combined with other letters, or
other marks according to any definite laws; and,
in the second place, so far are they from having
any definite signification or meaning attached

to them, that every chemist thinks himself at liberty to deal with them in this respect just as he pleases, according to his fancy. I wish to put a restriction upon that mode of dealing with the subject, and to bring my fellow-chemists and myself, when we have to deal with symbols, under some definite rules. Symbols are of two kinds. We may have symbols of things, and we may have symbols of operations. Symbols of operations are simply symbols of what we do to things. Take a popular case; ordinary language is an imperfect symbolic system, and here we have just these two kinds of symbols. A "dog" is the symbol of a thing, and "beating," "caning," "coaxing," and so on, are the symbols of operations, or of something which we may do to a dog. We have marks by which we express things, and marks by which we may express what we do to things. We might also have a third kind of symbol; we might have the symbol of an operation and a thing together.

Now before I commence my explanations, I should like to remove one or two popular errors upon this subject. I believe there is no error more ingrained in the popular mind than that the marks +, −, ×, =, are necessarily the symbols of adding, subtracting, multiplying, and identification or equalization; I mean that these marks are purely arithmetical symbols, and are to be used for purposes of arithmetic alone, and that in any other subject matter to which they are applied it is essential for us to give these symbols their arithmetical signification. If that were true, the application of symbols to the science of chemistry in any extended sense would simply be, from my point of view, an impossibility.

Perhaps I shall best illustrate this matter by giving you from another subject an example of the mode of constructing a symbol and of what we mean by a symbol. It is an example which will bring before you clearly how independent the application of symbols is of arithmetical meaning and interpretation. I say of arithmetical meaning not of arithmetical laws, which is another thing. In the ordinary geometrical

interpretation of algebra we denote by the mark
a the operation of conferring upon the unit of
length a certain specified length. To fix our
ideas let us take this length as three inches.
The mark *a* then, will thus stand for a
straight line, three inches in length. Now the
symbol + is what may be termed a directive
symbol and indicates to us the direction in
which the line *a* is to be drawn towards, let us
say, some specified point in the horizon. Hence,
if the line *AB*

$$A \underline{\qquad\qquad} B$$
$$a$$

be a line three inches long, *AB* will be properly
represented by the letter *a*, and + *a* will represent that line drawn from *A* to *B* ; and assuming, as I said, the symbol − to be similarly a
directive symbol, telling us to draw the line in
the opposite direction to that indicated by +,
− *a* will indicate a line three inches long drawn
from *A* to *C*

$$C \underline{\quad} A \underline{\quad} B$$
$$-a \qquad +a$$

B

Similarly, by the mark b we may represent a line five inches long, drawn in the same direction as a. Now, if we ask the meaning of the expression $a + b$ or $+ a + b$, the symbols inform us, putting BC as a line five inches long, that we are to commence by drawing as before from A to B the line a, and then to proceed to draw another line from B to C equal in length to b, as indicated below:

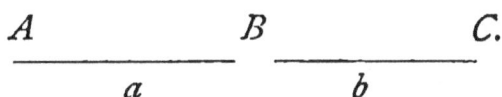

$$A \qquad\qquad B \qquad\qquad C.$$
$$\underline{} \quad \underline{}$$
$$a \qquad\qquad b$$

It follows that $a + b = b + a$ and $+ a + b = + (a + b)$, since it is indifferent as regards the total length and direction of the line, whether we commence by drawing the line a and continue by drawing b, or commence with b and then proceed to draw the line a; one peculiarity of this treatment of the subject, which is the ordinary geometrical application of algebra to geometry, being that we may always replace, without affecting the truth of the statement, the letters *a* and *b* by the arithmetical value of the length of the lines indicated by them.

We might also have argued thus : Let *a* be the operation performed upon a point by which a straight line three inches in length is generated. This operation is the transference of a point from one position to another without changing the direction of the transference. Again, let + be a directive symbol indicating the direction in which the transference occurs. We have then, referring to the figures above, + *a* as the symbol of the transference of a point from *A* to *B*, by which the line *AB* or *a*, is generated, and + *b* the symbol of the transference of a point from *B* to *C*, by which the line *BC* or *b*, is generated. A little consideration will show that the laws previously enunciated, $a + b = b + a$, $+ a + b = + (a + b)$, hold equally good with this interpretation as with that previously given. In this case also we can always substitute for the letter by which the line is represented, the number which expresses its length. It is, however, to be noticed that we cannot, by the instrumentality alone of the symbols hitherto employed, express lines drawn in any other direction than that indicated by the symbols + or −, namely, lines drawn in a specified direction and the opposite of that direction.

But another kind of algebraical geometry has been invented (what is termed double algebra), in which the symbols a, b, c, and so on, indicate to us not length alone but direction also, and are to be interpreted as the operations of conferring upon the unit of length, not only certain lengths, but certain lengths in any specified direction. So that, taking AB as a line three inches long, drawn in the direction indicated by the operation a, and AC as a line five inches long, drawn from the point a in the direction indicated by the operation b, the symbols a and b will indicate the lines AB, AC, as shown in the annexed figure:

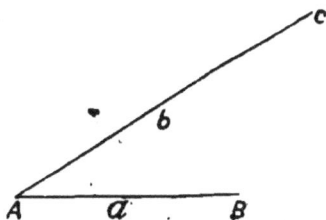

and the same principle of interpretation will prevail in the case of any number of symbols *a, b, c, d, e.*

It is to be observed that in this method lines are said to have the same direction which are parallel to one another. This method is termed Double Algebra, "from its meanings requiring us to consider space of two dimensions (or area), whereas all that ordinary algebra requires can be represented in space of one dimension (or length)." [1]

Let us now consider how, on these principles, the symbol $a + b$ is to be interpreted. *a* tells me to draw a line from the starting point, the line *AB*, three inches long. The symbol $+ b$ tells me to go on and draw at the termination *B* of the line *AB* the line *BC* (in the direction indicated by *b*) five inches long, which we may consider effected in the figure below. The direction of the line *b* being here assumed to be

[1] De Morgan, *Trigonometry and Double Algebra,* 1849, p. 117.

that of a line inclined to a at an angle of
35°.

Join AC. Now, I say that the line AC represents and is identical with the result of the algebraical sum of the operations a and b, that is, $a + b$, or, which is the same thing, $+ a + b$.

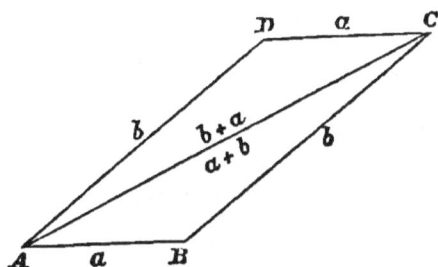

The reason of this statement may be thus given. Regarding a and b as the symbols of the operations of the transference of points by which the straight lines AB and BC are generated, the straight line AC is generated by the aggregate of these operations; the result being precisely the same in both as regards the direction and quantity of motion whether we transfer the point from A to B along the straight line AB and then by a second transference from B to C along the straight line BC, or transfer the point immediately from A to C along the straight line AC. This diagonal, however, is not equal in length to the sum of the sides of the parallelogram AC, but nevertheless this statement is correct; what we here denominate addition being truly not addition of magnitude to produce magnitude, but junction of effects to produce joint effect.[1]

Those persons, however, who consider it necessary that all algebraical symbols should admit of an arithmetical interpretation, must, if consistent, reject an algebra founded upon these principles.

We may note in passing, that these observations apply to the geometrical application of algebra alone. In the application of algebra to mechanics, for example, the diagonal actually represents, not only from the point of view of algebra, but also of arithmetic, the aggregate of

[1] De Morgan, *lib. cit.*, p. 118.

the forces represented by the sides of the parallelogram. Here is not the place to pursue this

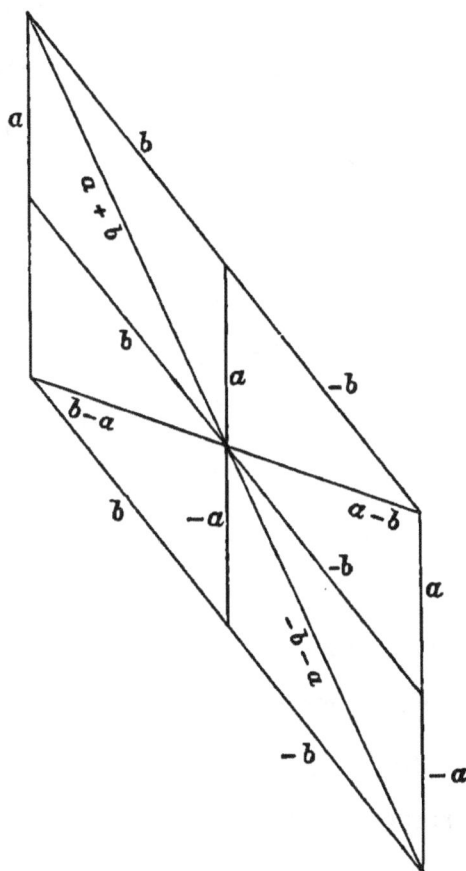

a b

$a+b$

b

a $-b$

$b-a$

b $-a$ $a-b$

$-b$ a

$-b-a$

$-b$ $-a$

subject, but the above diagram will convey all
the information in regard to it necessary for my
present purpose.

Having made these few observations in reference to symbols in general, let me proceed to
explain more precisely what I mean by a
chemical symbol. The object, I should say, of
the first part of this method (to which I must
refer you for fuller explanations) is mainly to
discover a proper system of symbols by which
we may express the units of chemical substances.
I may put this in another way, and say that
we wish to discover what is the nature and the
number of the operations by which chemical substances are made or constructed. That is the
first object of our method. I should, perhaps,
limit myself a little further, for I should say
that (in order to fix our ideas) before we begin
to consider such questions at all I shall conceive
of chemical substances as brought into the condition of perfect gases. The main reason of this
is the simplicity of the laws to which gaseous
compounds are subject, which simplicity was first
discovered by the great chemist, Gay-Lussac, and
which greatly facilitate the study of the question.
Of course, we may, if we please, deal with the

properties of the combinations of solids and
liquids, and regard the units of matter as existing in these forms. But here it is far more
difficult for us to arrive at any intelligible and
simple results; and therefore, before beginning to
think about the transformations of a chemical
substance, I, for my part, always conceive it as
brought into the condition of a gas. And to go
a little further, and to speak a little more defi-
nitely still, we shall always consider the chemical
substance as brought into the condition of a
gas, as the standard temperature of o degrees,
and at a pressure of 760 millimetres. The units
of all chemical substances are thus regarded
from the same point of view, without which no
comparison of these changes is possible. This is
the sort of chemical world with which we have
to deal, a world of gases.

First, let me give you the definition of a unit
of matter; for it is absolutely essential, before
we attempt to assign symbols to units of matter
to know precisely what we mean by the unit
which we are about to consider and symbolize.

That definition is of such great importance that I have had the words placed up before you in a diagram.

The unit of ponderable matter, is that portion of ponderable matter, which, in the condition of a perfect gas, at a temperature of o degrees, and at a pressure of 760 millimetres of mercury, occupies a space of 1,000 cubic centimetres.

From considering the unit of matter, I pass now to the consideration of a unit of another kind, and that is what I have termed the unit of space, which is the volume of 1,000 cubic centimetres of empty space. Now, we cannot work on this method until we have got hold of this unit of space, which is the subject on which the chemical operations by which the units of matter are constructed, are performed, and constitutes a fundamental conception peculiar to this calculus ; let us therefore endeavour clearly to understand what the unit of space means. Now, that there may be no doubt upon this point, I have brought you a very good image of the unit of space which is represented by this hollow cube with glass walls, and of the dimensions above assigned to the unit of space.

You must, however, go a step further. It is, indeed, the space of 1,000 cubic centimetres which is confined within these glass walls; but, before you can get at the unit of space, you must, by the process of imagination, or by the efforts of reason, divest this cube of glass of weight, and take out of it all the ponderable matter which it contains, and conceive the space within its walls divested of matter altogether. Now, this unit of space is fundamentally important to us, and I shall begin by giving it a mark to itself. The mark which I give to that unit of space is for certain good reasons which I will not explain now, but which I have fully given elsewhere, the mark 1. When you see that mark, it is to recall to your mind the matter contained in the unit of space. Now, what is that matter? Obviously, as there is no ponderable matter in it, that matter is *no* matter at all. The mark 1, therefore, is the symbol of the unit of no matter.

Perhaps, however, if I were to speak a little more precisely, I should say, for the benefit of those persons who may be more philosophically inclined, that the mark 1, from the point of view of operations, is to be defined as the symbol of the operation of taking the unit of space as it is. The symbol 1, therefore, tells us to take the unit of space as it is, and do nothing at all with it.

However, we have not to consider units of space, the consideration of which alone would lead us to very little, but we are going to consider the units of matter. Now, how are we to conceive of space becoming matter, or of matter getting into space—chemically, I mean? I shall think of this through the aid of an operation, and I shall define by a mark the operation by which this empty unit of space is turned into a unit of ponderable matter. For example, I will take x as such a mark. This is the mark of the operation by which the unit of space becomes a unit of ponderable matter of a certain specified kind and density.

Assuming, then, x as the symbol of such an operation; how are we to symbolize the performing of this operation upon the unit of space? I shall do this in the usual way in which the performance of operations on any given subject is symbolized, by writing the letter x before the symbol of the unit of space, thus: $x\,1$, and that indicates to me a unit of matter of a certain kind x, at 0° C. and at 760 millimetres pressure.

But we may be called on to represent a unit of matter double the density, but the same in kind as x. How is this to be effected? Having once conferred upon the unit of space this density x, we have then to perform the operation x a second time. Hence, to double the density we have only to write x again; thus, $x\,x\,1$, or $x^2\,1$. This will symbolize that we confer on the unit of space a certain density, and having done that, we confer that density on it again; that is, we make it double the density. Similarly $x\,x\,x\,1$, or $x^3\,1$, will mean that we give it three times the density, and so on. If you compare these operations with the symbols which express the densities, you will see that the symbols of the units of matter which we have thus constructed stand to the numbers which express the densities of that matter, in the same relation as numbers to their logarithms.

We will now take another kind of matter: y 1, y^2 1, y^3 1. These, again, would be the symbols of portions of ponderable matter which would be contained in this glass box at the pressure and temperature before-named of the kind indicated by y, and of the relative density indicated by the number of units of y. In these cases we have considered the construction of matter of one kind, but of different densities.

If we proceed further upon the same lines we come to consider the symbol of units of space containing two kinds of matter. Reasoning on the same principles as before, we have xy 1 as the symbol of the unit of space containing the matter of x, and also containing the matter of y; that is to say, having the density which is the sum of the densities of x and y. And, we can in this way symbolize also the unit of space filled with the matter x and y in various proportions.

You will see that there is a real analogy between the symbols which I am here employing, and the symbols which I used just now in my illustration derived from double algebra: we have the chemical symbol xy, the product of two chemical operations subject to the same algebraical laws as the arithmetical symbol $x\,y$, the product of two numbers, but with a totally different interpretation; and just as the symbols of double algebra indicate to us, not only the length of a line, but also its direction or position, so these chemical symbols indicate to us, not only the weight, but also the kind of matter. You are not to confound them with the numbers which express the densities, or the letters by which we might express those numbers; but they are, I say, symbols which express to us, at one and the same time, the nature of the matter and the density of the matter, having this double signification.

Before we go further, let me say a word about the nature of these operations x, y, and the like ; I am here symbolizing the unit of matter by the aid of the symbols of the operation by which the unit of matter is made. The question arises, what is that operation? The operation is one which, speaking with a certain degree of freedom, I may term a "packing" operation. It is the operation by which matter is "packed" into space, being, in fact, the operation with which every chemist is familiar under the name of combination, which is an operation precisely of this kind. But it is necessary for the comprehension of the methods of this calculus to enlarge our view of the nature of combination so as to include under this term (what is truly included under the same fundamental conception), not only the combination of matter with matter, but the combination also of matter with space.

We are getting thus at a definition of our unit in terms perhaps more in accordance with ordinary language. We will call the matter of x, A, and the matter of y, B; and the matter of

C

the unit of space o.   What, then, does x stand
for, considered from the point of view of combination ?   It is the operation of combining the
matter A with any substance which we please
to write after the symbol of the letter; and
y is the symbol of similarly combining the
matter B.   Now, if we write x before the symbol of the unit of space 1, thus: x 1, x 1 tells
us that we are to take the matter A and
combine it with the matter of the unit of
space, that is to say, to pack it into that
empty box which I have symbolized as 1, the
result being to constitute the matter A.   If,
having done that, I write y to it, thus: $x y$ 1,
this symbol tells me to take the matter B
and combine that also with the matter of the
unit of space.   If you do that, the result is
the matter of A combined with the matter of
B and the matter of the unit of space o.   That
is to say, those three things are combined together.
Do not imagine there is anything mysterious
about these terms.   They are operations about
which you think every day of your life; and,

if you want to think to any purpose about chemistry by means of symbols, you must embody in your symbols the very thing which you are thinking about, namely, the processes with which you have to deal.

If some curious person wishes to penetrate further still into the problem, and inquires in what the operation of combination consists, the only kind of reply I can make to him is to show him the result of combination and to explode, we will say, 2 vols. of hydrogen, and 1 vol. of oxygen, and exhibit to him the 2 vols. of water which is the result of this experiment. The combination, I say, of 2 vols. of hydrogen, and 1 vol. of oxygen, being merely the name given to the operation performed upon these quantities of these gases which produces this result. Various hypotheses, both metaphysical and atomic, have been framed to explain what combination consists in, but such hypotheses have not, at least in my judgment, thrown the slightest light upon the question.

The views of chemists as to the use of the

apposition of letters as the symbol of combination are of a very vague character. Berzelius, the author of our present notation, regarded the expression HCl as an abbreviated form for H + Cl. The late Sir John Herschel was very unwilling to admit the expression at all, and took the same view of it. The statements made on the subject in some of our chemical manuals are almost unmeaning. All that it is necessary to say on this point is that a system of chemical symbols which contains no distinctive symbol of combination, omits the most essential point to be considered, and that an indefinite symbolism to which no exact meaning is attached is necessarily of little value.

I must not seek to explain to you now the process or method by which we arrive at the symbols of chemical substances, that is to say, why I write the unit of hydrochloric acid as $a\chi$ and the unit of chlorine as $a\chi^2$. To explain the process on the board, and to do it any justice, would occupy far more time than is at my disposal, and it has been fully explained elsewhere. I will only ask you to allow me now simply to explain what we mean by the symbols of chemical substances in one or two special cases, and then to consider the general results to which this mode of representation conducts us.

As to the mode of constructing these symbols, it is quite a mistake to suppose that our symbols are the result of invention or hypothesis. They are based in the most absolute sense upon facts. We do not imagine or invent a symbol at all. We look for the symbol and find it. But where are we to look for the symbols of the operations by which units of matter are made? Plainly in the very facts of combination, to which I have just now referred. That is the source, and the only source open to us, whence to derive the symbol. The facts referred to in the case of gaseous combinations are such as these: 2 units of hydrochloric acid consist of the same ponderable matter, as 1 unit of hydrogen, and 1 unit of chlorine; 2 units of gaseous water consist of the same ponderable matter as 2 units of hydrogen and 1 unit of oxygen. Again, 2 units of ammonia consist of the same ponderable matter as 3 units of hydrogen and 1 unit of nitrogen. These are the facts, and chemistry supplies us with a large number of such facts. The method which I have ventured to give is merely a method of expressing these facts in the symbol of the substance. It is simply and purely, I say, a method of taking an equation expressing a chemical metamorphosis, and of embodying in the symbol certain facts of the equation. Through the facts of the equation we construct the symbols of the units of ponderable matter. We then take the symbols out of the equations, and thus separate and analyse the facts one from the other. It is an analysis of a peculiar kind.

I have constructed some tables expressive of the general nature of the conclusions at which we arrive through the aid of this method, as to the composition of these units of matter. I have had a good many of these symbols written out, for really it is easier for you, by looking at these tables, to see the general results which we arrive at by this method, than it would be for me to enter into a long explanation of the process. Here you see are the symbols of the chemical substances. We start with the symbol of the unit of space :—

Symbols of the Units of Chemical Substances.

Unit of Space	1
Hydrogen	a
Oxygen	ξ^2
Water	$a\xi$
Peroxide of Hydrogen . . .	$a\xi^2$
Sulphur	θ^2
Protosulphide of Hydrogen . .	$a\theta$
Bisulphide of Hydrogen . . .	$a\theta^2$
Sulphurous Anhydride . . .	$\theta\xi^2$
Sulphuric Anhydride	$\theta\xi^3$
Sulphurous Acid	$a\theta\xi^3$
Sulphuric Acid	$a\theta\xi^4$
Chlorine	$a\chi^2$

Hydrochloric Acid	$a\chi$
Hydrochlorous Acid	$a\chi\xi$
Chlorous Acid	$a\chi\xi^2$
Chlorosulphurous Acid . . .	$a\chi^2\theta\xi$
Hypochlorosulphurous Acid .	$a\chi\theta\xi^3$
Chlorosulphuric Acid	$a\chi^2\theta\xi^2$
Iodine	$a\omega^2$
Bromine	$a\beta^2$
Nitrogen	$a\nu^2$

In the next Table is another system of symbols, those of the combinations of carbon, hydrogen, and two or three other elements :—

Carbon	κ
Acetylene	$a\kappa^2$
Marsh Gas .	$a^2\kappa$
Olefiant Gas	$a^2\kappa^2$
Carbonic Oxide	$\kappa\xi$
Carbonic Acid	$\kappa\xi^2$
Alcohol	$a^3\kappa^2\xi$
Ether	$a^5\kappa^4\xi$
Glycol	$a^3\kappa^2\xi^2$
Glycerine	$a^4\kappa^3\xi$

Anhydrous Acetic Acid . .	$a^3\kappa\xi^3$
Tetrachloride of Carbon . . .	$a^2\chi^4\kappa^2$
Chloroform	$a^2\chi^3\kappa$
Chloracetic Acid	$a^2\chi\kappa^2\xi^2$
Trichloracetic Acid .	$a^2\chi^3\kappa^2\xi^2$
Chloride of Benzoyl . .	$a^3\chi\kappa^1\xi$
Cyanogen	$a\nu^2\kappa^2$
Hydrocyanic Acid . .	$a\nu\kappa$
Methylamine .	$a^2\nu\kappa$
Mercuric Ethide .	$a^5\kappa^4\delta$

You must regard these symbols as being, if I may so say, chemical equations turned into another form, and divested of a certain amount of superfluous and useless matter, which we do not want now to consider or think about. Nature does not supply us with the key-note to enable us to construct any one system of chemical symbols, necessarily true to the exclusion of every other system. Nature does not tell us absolutely—though I think she does tell us probably—how we are to proceed to construct such a system. In order to be able to construct a chemical system we must start with an hypothesis. As we go on constructing our symbols, our hypothesis, in so far as we prove it, approximates more and more to fact; but we must, at any rate, start with the assumption that we know one symbol. We may construct a complete chemical system from one symbol; and we may view all these symbols as the result of one hypothesis, combined with the facts given to us and supplied by the equations. Now, the hypothesis here made is that the symbol of the unit of hydrogen is expressed by one letter, *a.* That is my starting-point; and I should say that the symbols which you see in the tables, as indicating simple chemical operations, and expressed by one letter, are to be regarded as symbols of primary operations, that is to say, operations which you cannot resolve or decompose into any other operations by known methods.

They are symbols of primary operations; and when I say that the symbol of hydrogen can be expressed in chemical equations by one letter, I mean that in the changes and transformations of chemistry that unit of hydrogen is never broken up ; that it moves as a whole from system to system, and is never decomposed or resolved into parts. Hydrogen is constructed at once, by one operation. Imagine yourself witnessing the formation of hydrogen. To form some substances you want many operations ; but to form hydrogen you want only one operation. That [striking a blow on the glass model of the unity of space] represents the formation of hydrogen—*one* operation. It is one act. If we could witness chemical transformations, and nature should become vocal to us, and indicate each combination as it occurred by a musical note, that [again striking a blow] is what you would hear when hydrogen was formed. Now, as we go on we come to much more complex substances. Let us take oxygen. This is a substance very different indeed from hydrogen in its chemical properties ; and as you can conceive of the unit of hydrogen being made at once by one operation, so I say that it is impossible for you to conceive of the unit of oxygen being made by less than two operations. To return to our metaphor. When you take water and decompose it, so that oxygen is formed, you ought to hear two notes. That is what I mean when in this language I say that oxygen is made by two operations. Again, the unit of water is made by two operations like the unit of oxygen; but it differs from the unit of oxygen in this respect, that one of those operations is the same as that by which hydrogen is made. That is to say, in the operation by which water was formed you would hear two notes, one different from the other, a, ξ.

The symbol of chlorine is $a\chi^2$. Chlorine, from this point of view, is to be conceived as made up by three operations. You are to hear χ, χ, a. One of these operations is the same as that by which hydrogen is made, and the other is an operation peculiar to chlorine itself, namely, χ. Again, a unit of hydrochloric acid is to be conceived of as made by two operations, a and χ.

To go one step further: let me refer you to this Table :—

Nitrogen	$a\nu^2$
Ammonia	$a^2\nu$

Protoxide of Nitrogen . . . $a\nu^2\xi$

Nitrous Acid . . . . . . $a\nu\xi^2$

Nitric Acid . . . . . . $a\nu\xi^3$

Phosphorus . . . . . . $a^2\phi^4$

Phosphide of Hydrogen . . $a^2\phi$

Hypophosphorous Acid . $a^2\phi\xi_5^2$

Orthophosphoric Acid . . . $a^2\phi\xi^4$

Terchloride of Phosphorus . $a^2\phi\chi^3$

Pentachloride of Phosphorus . $a^3\phi\chi^5$

Nitrogen is to be conceived of here as made by three operations, ν, ν, a. In the formation of the unit of ammonia also three operations concur ; one of them being one of the operations of nitrogen, ν, and the other two being the operations by which hydrogen is formed, a.

I must not enter into further details upon this subject, but I have little doubt that, with this explanation, you will readily appreciate the meaning of the symbols which are written up before you. You will see that, by following this process of taking the facts of the equations and turning them into the language of symbols, we arrive at a peculiar view as to the nature of matter, which view is embodied in those symbols.

Now, as to the view of the nature of the elemental bodies which is here indicated ; for that, perhaps, will occur to many persons as the most important point to be considered, for, seeing that it is out of these elemental bodies that everything else is made, and that into them all things are capable of being resolved ; the view which we take of these bodies gives us implicitly the view which we are to take of the composition of every other body whatever. To understand this it is only necessary to appreciate the view which is here given of the nature of the elements themselves, and everything else follows from that. We are led to the following singular results, — that, speaking generally, there are, perhaps, four—and certainly, at least three—fundamentally distinct classes of elements.

First of all, the elements, the units of which are made by one individual operation. These bodies are represented to us by mercury and hydrogen. To this class also probably belong such elements as zinc, cadmium, and tin; but we cannot speak with great confidence on that point.

Secondly, we have a class of, so to say, double elements formed by two similar operations; these are such as oxygen, ξ^2, sulphur, θ^2, selenium, λ^2. Carbon we are not certain about; it belongs, in all probability, to the first or second class, we do not quite know which; but I have symbolized it as κ^2.

But we have another and a very large class— perhaps the largest of all the groups of the elements—and we may take the elements chlorine and nitrogen as representatives of it. Here is the symbol of the element chlorine, $a\chi^2$; here is nitrogen, $a\nu^2$; here is iodine, $a\omega^2$, and so on. You will see that the symbols of these elements occupy a certain intermediate portion between the group of elements, a, δ, ζ, &c., and the group of elements ξ^2, θ^2, λ^2, &c. We have many compound substances which are in every way analogous to this group of elements—analogous as to their properties, analogous as to their symbols. Of this class we have a most interesting and striking example in the peroxide of hydrogen; which is symbolized here as $a\xi^2$. You see the peroxide of hydrogen is really to be regarded as the combination of one unit of the element hydrogen with one unit of oxygen—which things really exist,— just as the element chlorine may be regarded as a combination of the unit of hydrogen a with the unit of a substance which does not exist, and which I have symbolized as χ^2. The unit of nitrogen is to be regarded as similarly composed, $a\nu^2$. We may regard it as a combined with the unknown element ν.

There is one question which must occur to every one, the explanation of which is of fundamental importance to the comprehension of this system. You may ask me, " What reality do you attach to these symbols? When you call chlorine $a\chi^2$; nitrogen, $a\nu^2$; oxygen, ξ^2; do you make the hypothesis that there are certain real bits of matter actually, or even possibly, existing capable of being brought to the lecture-room and exhibited on the table—bits of matter which you represent by a, χ, ν, and the like; do you mean this? or do you mean that these things do not exist, that they are the mere creation of your imagination, fictions, illusions? We like Dalton," perhaps you may say to me—"we like Dalton far better than we do you; for Dalton made no such claims on our imagination. He, at any rate, was intelligible, and dealt with realities, or possible realities, alone. He showed us the elemental matter of which all substances are made; and even in his atoms Dalton dealt with what he believed to be realities. Neither he nor we indeed have ever seen these things; yet, nevertheless, we most perfectly believe them to exist. To impress their reality upon the mind Dalton drew pictures of them, and made bits of wood to represent them; by which he certainly went so far as to express his belief that they were real material things of definite form. Now can you also do this for us? can you show us the matter of which these elements, ξ, χ, ω, ν . . . . consist? Will you take a piece of chalk and draw upon that board some picture, or figure, or diagram to render clear to us what these things are?" To these perplexing questions I cannot give a direct answer. The symbol of a simple weight is not necessarily the symbol of a real thing. I have never assumed it to be so, and I have never attempted to prove it to be so. I cannot draw a picture, or represent by a model the structure of a thing which is not real. On the other hand, these symbols are not the creation of my imagination. I did not invent them; I only found them in the course of an analytical process. It is, therefore, equally untrue to speak of them as unreal, for I do not know this to be the case. Now, a thing which is neither real nor unreal, but may be either, is that which I here term an "ideal" thing; and for this reason I speak of the factors by which in this calculus the symbols of the units of matter are expressed as "Ideal" factors, and in this they essentially differ from the corresponding representations, afforded by the atomic theory, which, being a theory or hypothesis as to the constitution of matter, deals with realities alone. The essential point is that in this calculus it is not necessary to pronounce any further opinion upon this question, for it is proved that, so far as all analytical ends are concerned in considering and reasoning upon the problems of chemistry by means of analytical processes, it is totally unnecessary to raise this question, and we may confidently deal with the ideal factor as with real factors, satisfied that we cannot be led into error by so doing. The ideal weight is a thing which may exist or may not exist, as an external reality, but for those purposes of reasoning with which we are here concerned it satisfies all the analytical conditions supplied to us by chemical equations, and we are bound to accept it as a member of the general system of symbols.

I will venture to give you an illustration on this subject which was suggested to me by some remarks of Professor G. G. Stokes, with whom I have had the great advantage of discussing several of these abstruse questions. The following statement is a mathematical truth invariably admitted: *every straight line cuts every conic section in two points.* This assertion may be considered to correspond to my statement that the unit of every chemical substance is compounded

of an integral number of simple weights. But
you say, "Do you mean that every straight line
cuts every conic section in two real points? If
so, you should be able to explain by means
of a geometrical diagram how and where it
cuts it." To this I reply, that my assertion
cannot be represented at all by means of a
geometrical diagram : that the statement is not
a geometrical but an algebraical truth. I never
said that the straight line *really* cut the conic
section at all. I said that it *cut* the conic
section, and I will supplement my previous
statement by saying that every straight line
cuts every conic section in two points, which
are real, coincident, or imaginary. Similarly,
I say that every unit of matter is made up of
an integral number of simple weights not necessarily real, but which may be either real or
imaginary, although we have not the data to
determine to which class they belong. Now, as
the statement that every straight line cuts every
conic section in two points is not a geometrical
but an analytical, or symbolical truth, and we

cannot, speaking generally, and without reference to a particular case, draw a geometrical diagram indicating these points, so also in the simple weights of chemistry we cannot draw on the board visible pictures to represent them. This is possible in the case of the Daltonian atom. But the only possible representation of the simple weights of this calculus is the symbols by which they are expressed in the analytical system of which they are members, and any other representation must necessarily mislead.

Now, although it is essential carefully to discriminate between the symbolical expression employed in this calculus, and any physical hypothesis based on this analytical expression, yet we cannot altogether disregard the alternative that the portions of matter symbolized by a, χ, ξ, v . . . . may be real physical existences. This hypothesis cannot be established by means of any symbolical calculus, for we cannot infer because the symbol of chlorine may be expressed in every chemical operation by the three letters a, χ, χ, that the matter of chlorine is made up of three real distinct bits of matter into which in chemical transformation it is resolved, and which are capable of a real and independent existence; but, nevertheless, there are very forcible reasons which (when once we are in possession of this symbolical system) lead us to suspect that chemical substances are really composed of a primitive system of elemental bodies, analogous in their general nature to our present elements, some of which we possess, but of which we possess only a few. I will take the case of peroxide of hydrogen. Neglecting oxygen and a great class of oxygenated combinations, I will suppose for the moment that I have these combinations in my hand—hydrogen, water, peroxide of hydrogen, and certain other substances which I could specify. If I were to apply my method to finding the symbol of peroxide of hydrogen, not regarding the oxygen at all, the symbol at which we should arrive for peroxide of hydrogen is $a\xi^2$. Then the same question would arise about peroxide of hydrogen as now arises about chlorine, namely, whether the bit of matter represented by ξ were real or imaginary. In the case of peroxide of hydrogen we have, however, really succeeded in separating the elements which it contains, and this fact among others leads us to the suspicion that some of these bodies which we speak of as elements may in fact be compounds. In short, we are led, through our method, to a certain physical hypothesis as to the origin and causes of chemical phenomena.

Now, what I am going to suggest you must consider to be put before you with reservation, but we may conceive, that, in remote time, or in remote space, there did exist formerly, or possibly do exist now, certain simpler forms of matter than we find on the surface of our globe —a, χ, ξ, ν, and so on—I say, we may at least conceive of, or imagine, the existence, in time and space, of these simpler forms of being, of which we have some records remaining to us in such elements as hydrogen and mercury. We may consider that in remote ages the temperature of matter was much higher than it is now, and that these other things existed then in the state of perfect gases—separate existences—uncombined. This is the furthest barrier to which in the way of analysis theory can reach. Beyond, all is conjecture. There may be something further, but if so, we have no suspicion of it from the facts of the science. We may, then, conceive that the temperature began to fall and these things to combine with one another and to enter into new forms of existence, appropriate to the circumstances in which they were placed. We may suppose that at this time water ($a\xi$), hydrochloric acid ($a\chi$), and many other bodies began to exist. We may further consider that, as the temperature went on falling, certain forms of matter became more permanent and more stable, to the exclusion of other forms. We have evidence on the surface of our globe itself, of the permanence of certain forms of matter to the exclusion of others. We may conceive of this process of the lowering of the temperature going on, so that these substances, $a\chi^2$, and av^2, when once formed, could never be decomposed—in fact, that the resolution of these bodies into their component elements could never occur again. You would then have something of our present system of things. You might further imagine that it would be possible, on looking carefully at chemical equations, and minutely studying them, to recover from the equations the record of the truths which were buried and preserved in the equations; and some analyst might come and say, "These equations are *only* consistent with this hypothesis, that chlorine is composed of a and χ^2," or, at least, it might be said that the equations *are* consistent with that hypothesis, for I do not want to go further than that. We can conceive, I say, of such a state of things. Now, this is not purely an imagination, for when we look upon the surface of our globe, we have, as I said before, actual evidence of similar changes in nature. We talk of the elemental bodies as though they were existing things; but where are they? We have oxygen, nitrogen, sulphur, certain metals, and certain bodies which we could specify, but what has become of the others? Where is hydrogen? Where is chlorine? Where is fluorine? Where are these things? They are locked up in combination in such a way that it is only within the last hundred years that the art of the chemist has revealed them to mankind. Now, if in our globe there had been more hydrogen—if there had been an excess of hydrogen present in the matter from which our globe was made—and if we suppose it to be true that the gases condense in the solid matter of our globe, we cannot doubt that the whole of the free oxygen would have been carried away from our planet, and that we should have had simply oxygen stored up in the form of water. We should have had water, but no oxygen at all; the hydrogen would have combined with it and carried it all away.

When we look at some of the facts which have been revealed to us, by the extraordinary analyses which have been made of the matter of distant worlds and nebulæ, by means of the spectroscope, it does not seem incredible to me that there may even be evidence, some day, of the independent existence of such things as χ and ν. We know that Dr. Miller and Mr. Huggins saw a most wonderful hydrogen combustion—at least what they imagined to be a hydrogen combustion—taking place in a variable star. Now this hydrogen combustion might be actually hydrogen combining with these unknown elements, and carrying them away in the form of chlorine, nitrogen, and the like. One of the nebulæ examined by Dr. Miller and Mr. Huggins afforded them the spectrum of an ignited gas, and in the spectrum of this nebula they saw one of the lines of nitrogen alone. This suggested to them that the line might have been produced by one of the elements of nitrogen. That might have been the element, ν. This as yet is a mere suggestion, but it seems to me eminently probable that if we follow up the subject we may from this source have one day revealed to us, independent evidence of the existence of these elements in the sun or stars. (See Note A.)

Let me, in conclusion, make one or two

observations upon a point which must occur to every chemist who has studied this method. If we had not taken a as the symbol of hydrogen, but had started with a different hypothesis, namely, that the symbol of hydrogen was a'^2, we should have arrived at a different symbolic system analogous in its form to our present system. We should have hydrogen as a'^2, water as $a'^2\xi$, and so on. In fact, we should have been led to develop a system different from that which I have brought before you.

In the following Table are given a few examples of symbols constructed on this hypothesis :—

Hydrogen	a'^2
Chlorine.	χ'^2
Hydrochloric Acid	$a'\chi'$
Hydrochlorous Acid	$a'\chi'\xi$
Chlorosulphurous Acid . . .	$\chi'^2\theta\xi$
Hydrochlorosulphurous Acid .	$a'\chi'\theta\xi^3$
Iodine	ω'^2
Nitrogen	ν'^2
Acetylene	$a'^2\kappa^2$
Marsh Gas	$a'^4\kappa$

Cyanogen	$\nu'^2\kappa^2$
Hydrocyanic Acid	$a'\nu'\kappa$
Ammonia	$a''^3\nu'$
Methylamine	$a''^3\nu'\kappa$

You may with reason ask me, "Why do you prefer one of these systems to the other? or do you prefer it? or what view do you take of that question?" Let me say, in the first place, that I cannot as yet give a complete answer to this question. For, I have not placed before you and others the ideas upon which a judgment can properly be formed upon it.[1]

I will, however, make one remark which will be sufficient to convince those who have so far followed me of the essential difference between the two systems. On comparing the second system, "system a''^2," with the first system, "system a," it will be seen that we may always, by a mere process of substitution, pass from the former to the latter, that is to say, every combination of

[1] This has since been done in Part II. of this Calculus. I refer especially to the discussion contained in it as to the origin of the law of even numbers.

the latter system will have its counterpart in
the former—the combinations being expressed in
the two cases respectively by positive and integral
members of the prime factors of the systems:
but it is not true that every combination of the
former system will have a counterpart in the
latter, or can be expressed by the prime factors
of that system; thus, for example, the combination $\nu'\xi$, which is a combination found in the
system a'^2, has no counterpart in the system a,
and cannot be expressed in it. The system a'^2
is therefore more comprehensive than the system a.
This observation disposes at once of the remarks
of those critics who maintain that because we can
pass by a simple process of translation from the
system a'^2 to the system a, these systems are to
be regarded as meaning the same thing, it being
perfectly indifferent to which we adhere. Such
persons are really in the position of those wise-
acres who maintain that because all A is B all
B is A. When we have to select between two such
hypotheses, the more restricted hypothesis, which
in this case is system a, is always to be preferred.

The reason of this restriction is that system *a* excludes all those combinations which do not satisfy the law of even numbers, of which the system a'^2 takes no notice. At this point I must leave the subject for fuller consideration hereafter.

———————

NOTE A.—Since this Lecture was delivered, further researches have been made in this direction, and in an article by myself in the *Philosophical Magazine* of June, 1879, the following passage occurs :—

It is a significant fact that a very large proportion of the class of elements which I have termed composite elements have not been found in the sun.

In reply to inquiries on my part, Mr. W. Huggins writes to me thus :—

"So far as I know, nitrogen, phosphorus, arsenic, antimony, boron, chlorine, iodine, bromine, have *not* been found in the sun. In one paper Lockyer *suspects* iodine. Dr. Miller and I found coincidence of three lines of antimony, with three lines in Aldebaran. Though this observation would show considerable probability of antimony in this star, I do not think the spectroscope (two dense prisms of flint glass) was sufficiently powerful to make its existence there certain. In the case of nitrogen, no coincidence was observed in any of the stars. In my paper in the *Transactions of the Royal Society*, on Spectra of Nebulæ, I show coincidence of principal line with the strong line in spectrum of nitrogen. Now, this line of nitrogen is a double one;

and I was not at first able to be certain if the line in the nebula was similarly double. Subsequently, with the powerful spectroscope I used for the motions of the stars, I was able to make a *certain* determination of this point (*Proceedings R. S.*, 1872, p. 385). I found the line in the nebula *single* and *coincident* with the middle of the less refrangible of the components of the double line.

Nitrogen	Red
‖	
Nebula	|

I say 'middle,' because the line in the nebula is narrower and more defined than either of the two lines forming the double line. I made experiments to see if, under any conditions of pressure and temperature, the more refrangible of the two lines fades out, so as to leave only the one with which the line in the nebula is incident. I did not succeed. So the matter stands : Is nitrogen compound ? Are there any conditions under which the one line only appears ? Has the line in the nebula no connection with nitrogen further than being sensibly of the same refrangibility?"

Now we must either consider that the matter of these elements, so abundant on the earth, does not exist in the sun or stars (which is not probable), or that they have passed into forms of combination in which they cannot be recognised by the spectroscope (which is also hardly admissible at that elevated temperature), or that they have been decomposed.—*Philosophical Magazine*, 1879, p. 130.

LONDON : R. CLAY, SONS, AND TAYLOR, PRINTERS.